NATURAL DISASTERS
AN IMAGINATION LIBRARY SERIES

PLAGUES

by Victor Gentle and Janet Perry

Gareth Stevens Publishing
A WORLD ALMANAC EDUCATION GROUP COMPANY

J

Please visit our web site at: www.garethstevens.com
For a free color catalog describing Gareth Stevens' list of high-quality books and
multimedia programs, call 1-800-542-2595 (USA) or 1-800-461-9120 (Canada).
Gareth Stevens Publishing's Fax: (414) 332-3567.

Library of Congress Cataloging-in-Publication Data

Gentle, Victor.
    Plagues / by Victor Gentle and Janet Perry.
        p. cm. — (Natural disasters: an imagination library series)
    Includes bibliographical references and index.
    ISBN 0-8368-2835-6 (lib. bdg.)
    1. Epidemics—Juvenile literature. 2. Animal introduction—Juvenile
literature. 3. Plant invasions—Juvenile literature. [1. Epidemics. 2. Animal
introduction. 3. Plant invasions.] I. Perry, Janet, 1960- II. Title. III. Series.
    RA653.5.G46   2001
    614.4—dc21   00-051623

First published in 2001 by
Gareth Stevens Publishing
A World Almanac Education Group Company
330 West Olive Street, Suite 100
Milwaukee, WI 53212 USA

Text: Victor Gentle and Janet Perry
Page layout: Victor Gentle, Janet Perry, and Joel Bucaro
Cover design: Joel Bucaro
Series editors: Mary Dykstra, Katherine Meitner
Picture researcher: Diane Laska-Swanke

Photo credits: Cover, p. 17 © Peter Johnson/CORBIS; pp. 5, 15 © C. Andrew Henley/Larus;
p. 7 © Historical Picture Archive/CORBIS; p. 9 © Arthur M. Siegelman/Visuals Unlimited;
pp. 11 (both), 21 © AP/Wide World Photos; p. 13 © Rob & Ann Simpson/Visuals Unlimited;
p. 19 © Gary Meszaros/Visuals Unlimited

Printed in the United States of America

1 2 3 4 5 6 7 8 9 05 04 03 02 01

Front cover: *Africa has a problem. Swarms of queleas devour its crops,
causing many people to starve. Some Africans have started eating the
birds. Are the people, or the queleas, a bigger threat to the other?*

TABLE OF CONTENTS

Words that appear in the glossary are printed in **boldface** type the first time they occur in the text.

TOO MANY, TOO FAST

You know what it's like when boring adults visit your house. They take over your room and eat all the cookies. You have to pretend you're interested in the questions they ask, the games they want to play, and the stories they tell. Like most annoying things, they eventually go away and your house and life return to normal.

A plague is a little like that. Except with a plague, you should multiply the "guests" by about a zillion. They might eat *all* your food, not just the fun food. You might end up getting very sick, instead of just annoyed. To top it off, these guests never go away until it's much too late to return to normal.

*A plague for a plague. Two dozen rabbits were set free in Australia in 1859. Soon, millions of them were eating up all the crops and plants. This rabbit is infected with a **virus** used to control the plague of rabbits.*

A PLAGUE ON YOU!

You may have heard stories about plagues like the one in the Bible about Moses. Moses asked the pharaoh, or leader of Egypt, to free the Jewish slaves.

When the pharaoh refused, ten plagues hit the Egyptians. First, the river ran with blood, killing all the fish. Second, frogs covered the land. Third and fourth, lice, then flies, tortured the people and animals. Fifth, the cattle died. Sixth, everyone got sores on their bodies. Seventh, hail destroyed the crops. Eighth, locusts ate every plant that was left by the storms. Ninth, three days of darkness fell over Egypt. Tenth, all of the first-born Egyptian sons died. The pharaoh finally freed the slaves. At last, the plagues stopped.

Slimed! This picture shows frogs plaguing Egypt. People believed frogs were evil creatures that spread disease. Today, a plague like this only seems a little squishy.

IT HURT A FLEA — AND US!

The plague that gives all plagues a bad name is the **bubonic plague**. In the 1300s, it stowed away inside of the fleas that gave it to the rats that gave it to the sailors that traveled on the ships from the Far East to Europe. Then, the sailors spread it to people all over Europe.

Bubonic plague killed millions of people — entire families in a week, and whole towns in a few months. People were terrified to leave home because they were afraid of catching the disease. Farmers couldn't harvest their crops or care for their animals. People and cattle starved and died. Bodies piled up in the streets because no one could bury them fast enough!

Bubonic plague is caused by a virus called Yersina pestis, which is passed along by fleas when they suck blood from rats, mice, and other mammals — like humans.

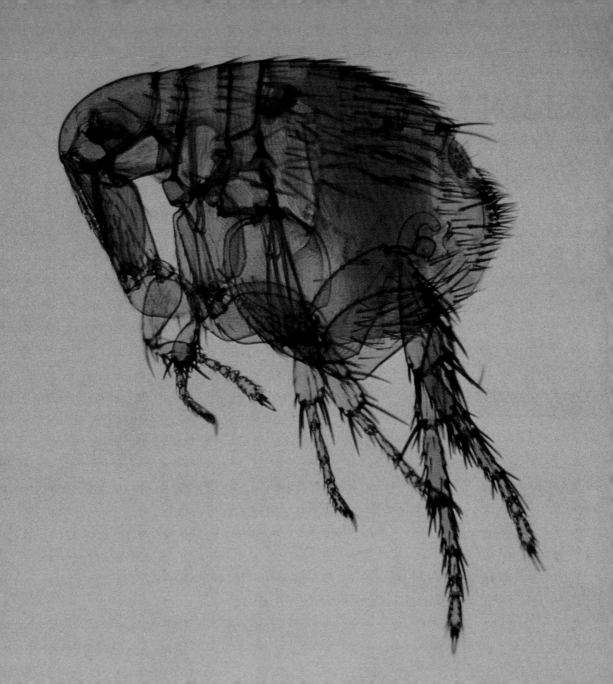

PEOPLE PACKING POXES

Other diseases have traveled with explorers and invaders. As Europeans journeyed around the world in the 1500s and 1600s, the smallpox virus rode along inside them.

Native peoples of the Americas and Polynesia did not have resistance to smallpox, so whole tribes got sick and were wiped out.

Now we understand how **epidemics** like bubonic plague and smallpox work. As a result, travelers are asked to get shots or take along medicine to keep them from carrying diseases from place to place. Of course, there are diseases that we have not discovered yet.

Asian longhorned beetles traveled with Japanese imports to Illinois. After arriving in the United States, the beetles tunneled into trees, killing them from the inside out.

MUCH TOO MUCH

Plagues often strike quickly and mysteriously, making people wonder why something so terrible has happened. The truth is, most plagues are started accidentally — by *us*.

When we send things from place to place, animals, plants, and germs sometimes hitch a ride in the crates, barrels, and suitcases that travel by air, sea, and land. When these **organisms** arrive in new places, they may kill off the organisms that already live there, upsetting the balance of nature in that area.

Entire **ecosystems** are changed — usually forever.

Zebra mussels entirely changed the Great Lakes ecosystem in North America. They smother the native clams and eat the clams' food. They also clean up human pollution.

Hate It! Warts and All!

Australians are losing a war caused by an invasion — and they *brought* the invaders into their country! In the 1800s, Australian farmers planted sugar cane to make money. Cane beetles came with the sugar cane and devoured the crop. So farmers brought South American toads to eat the beetles. But the toads ate many native Australian animals. Worse, the toads are poisonous. **Predators** won't eat them! OOPS!!!

Cane toads are ruining the Australian ecosystem. Yet in their native land, the toads are nearly an endangered species. Cane toads, go home!!!

Aussie scientists found a South American virus that kills cane toads, but it's too dangerous. This blue-green algae is suffocating the invaders. Maybe the swamp has a better plan!

CAN'T BEAT 'EM? EAT 'EM!

In Africa, some farmers are making a mistake that causes the destruction of their own crops. They plant grains that are easy winter food for small birds called **red-billed queleas**. Instead of migrating at harvest time, which is what queleas usually do, they stay year-round. So the queleas eat millet and oats, and the people go hungry.

Other African farmers have tried planting **maize** because queleas don't like it. In the deserts of Africa, there is little water. Often there are long periods without rain, and the corn dies. Then the people living in the region starve.

Still other Africans hunt queleas. Instead of trying to beat this "plague," they eat it instead!

Queleas swarm like bees over African crops. In some countries, people eat the critters that plague them. In Mexico and Thailand, people eat grasshoppers and locusts like french fries!

RIDING THE LIFE CYCLE

Inside certain ecosystems, what might seem like a plague is actually a normal **life cycle**.

In North America, insects called seventeen-year cicadas live as grubs for (you guessed it!) seventeen years. They crawl out of the ground in *huge* numbers. The winged adults soon cover tree trunks, grass, and streets. There are so many, you can't avoid stepping on them. It seems like a plague, but it's not.

Cicadas are members of their ecosystem, they don't destroy it. Birds and rodents love to eat cicadas. Cicadas may control weeds. By coming out rarely and in the *millions*, they can survive being eaten by the *thousands*. Their life cycle protects their chances of survival as a **species**.

Cicadas at dinnertime. Sections of fully grown trees will be destroyed as cicadas emerge for a day, eat, mate, and then die.

CONSTANT CHANGE

Are plagues ever good? It depends on the plague. Some people eat the insects and animals that "plague" them.

For us, a plague can be a very painful signal to change the way we do things. For example, we need to be more careful about the souvenirs we bring home from our travels. We also need to be careful how we "fix" the things we may not like inside an ecosystem. Remember what happened with the cane toads in Australia!

If a damaged ecosystem could talk, it might tell us that *we* are the plague!

Plagues can be controlled by watching world travelers carefully and keeping pests and germs out. This beagle is on duty, sniffing out anything that might cause plagues.

MORE TO READ AND VIEW

Books (Nonfiction) *Atlas of Earth.* Alexa Stace (Gareth Stevens)
Cicada Sing-Song. Densey Cline (Gareth Stevens)
Disaster! (Catastrophes That Shook the World). Richard Bonson
and Richard Platt (Dorling Kindersley)
The Nature and Science of Autumn. Jane Burton and Kim Taylor
(Gareth Stevens)
The Nature and Science of Spring. Jane Burton and Kim Taylor
(Gareth Stevens)
The Nature and Science of Summer. Jane Burton and Kim Taylor
(Gareth Stevens)
The Nature and Science of Winter. Jane Burton and Kim Taylor
(Gareth Stevens)
The Science of Insects. Janice Parker (Gareth Stevens)
Why Do We Have Different Seasons? (Ask Isaac Asimov).
Isaac Asimov (Gareth Stevens)
Wonderworks of Nature (series). Jenny Wood (Gareth Stevens)
World Almanac for Kids. Elaine Israel (World Almanac Books)

Books (Activity) *175 Amazing Nature Experiments.* Rosie Harlow, Gareth Morgan,
and Kuo Kang Chen (Random House)

Books (Fiction) *A Parcel of Patterns.* Jill Paton Walsh (Farrar Strauss Giroux)
Shakespeare's Scribe. Gary L. Blackwood (Dutton)

Videos (Nonfiction) *National Geographic's The Invisible World.* (National Geographic)

WEB SITES

If you have your own computer and Internet access, great! If not, most libraries have Internet access. The Internet changes every day, and web sites come and go. We believe the sites we recommend here are likely to last and give the best and most appropriate links for our readers to pursue their interest in plagues, critters that are out of control, and the kooky ways ecosystems work.

www.ajkids.com

Ask Jeeves Kids. It's a great research tool. Some questions to try out in Ask Jeeves Kids:
What is the best way to fight a plague?
What epidemics are predicted for this year?

You can also just type in words and phrases with "?" at the end, for example:
Australian Rabbits? Bacteria?

You can also enter the word "cicada" and get a box that will allow you to hear the sound of a cicada.

www.hhs.gov/kids

This is the U.S. Department of Health and Human Services site for kids. Click on Hey Kids or Calling All Students to learn interesting facts about toxic substances in the environment. Or, if you'd rather discover all the different diseases that exist in the environment, click on Environmental Diseases from A to Z.

www.commtechlab.msu.edu/sites/ dlc-me/zoo

Visit the Microbe Zoo to see all the critters that live in the invisible worlds around us, on us, and in us!

www.education-world.com/science/elem/ habitat_k_8.shtml

Go to Education World to find out how habitats work.

www.yucky.com

Yucky.com. Visit one of the most disgusting places on the web! Go to Yucky.com to ask Wendell the Worm all the questions you have about diseases and icky plagues. Click on colds and flu to find out how we catch stuff like that; click on yucky animals to learn about really gross critters; click on your body to find out how the most sickening thing about getting sick is what your body does to get and stay well!

www.affa.gov/au/docs/animalplanthealth/ plague

See how big a locust is, what it looks like at all three stages of its life, what it eats, where it can be found, and what we do to it to keep it from eating our food.

www.ex.ac.uk/bugclub

Join the Bug Club and find out the latest news about all kinds of insects, and whether they are pests or not.

GLOSSARY

You can find these words on the pages listed. Reading a word in a sentence helps you understand it even better.

bubonic plague (boo-BON-ik PLAYG) — an illness that makes some glands turn black, causing pain and even death 8, 10

ecosystem (E-ko-sis-tum) — a place with a certain weather pattern and organisms that live together, adapted to the place, its climate, and each other 12, 14, 18, 20

epidemics (eh-pih-DEM-ix) — plagues that spread quickly, causing panic, sickness, and sometimes death 10

life cycle (LIFE SI-kul) — events during the life of an organism or ecosystem 18

maize (MAYZ) — corn (sweet corn or Indian corn) 16

organism (OR-gun-is-m) — a living creature, such as an animal, plant, fungus, or germ 12

predators (PREH-duh-torz) — animals that hunt and eat other animals 14

red-billed quelea (RED bild KWEE-lee-uh) — a small, brownish bird with a red beak that lives in grasslands. It travels in large flocks, eating grains and termites 2, 16

species (SPEE-sheez) — a grouping of organisms 18

virus (VI-rus) — a tiny organism that lives off of a "host" organism. Viruses can cause diseases and eventually lead to epidemics 4, 8, 10, 14

INDEX

O/1. 9/02 2-15-05